Photochemical Smog

Dr. Hemant Pathak

ISBN: 1541065689
ISBN-13: 978-1541065680

DEDICATION

Dedicated to Shri Sainath Maharaj the all omnipotent
of world the most merciful.

CONTENTS

Foreword

Photochemical Smog was first uncovered in the 1950s with observations of its harmful effects on Nonliving beings in the vicinity of Los Angeles.

In India, roughly One sixth of the world's population lives in dense areas. There are now more than 45 cities with a population of over 1 million inhabitants. This tendency toward urbanization is running to continue.

Millions of people are being exposed to harmful levels of air pollutants caused mainly by emissions from industrial and domestic activities and from motor vehicles involving the combustion of fossil fuels.

Reduced visibility, damage to monuments and buildings, and many other such consequences indirectly affect our quality of life, ultimately damage to ecological systems. This Book represent state of knowledge regarding the nature and effects of photochemical pollution and draws some conclusions concerning the setting of goals for oxidant concentrations and the most effective strategies for smog control.

The nature of Photochemical Smog is explored with some emphasis on the limitations of ozone as an indicator of pollution. This book provides an essential guide to researchers, it offers: affects of Photochemical Smog on current Society.

Simply explained, Photochemical Smog is an important book bringing together diverse viewpoints from Environmentalist, state agencies and regulators, for all who wish to save Earth with quality Life.

Dr. Hemant Pathak

M.Sc. (Gold medalist), Ph. D.

Assistant Professor of Engineering Chemistry

Indira Gandhi Govt. Engineering college,

Sagar, MP, India

Acronyms and Symbol

CNG   Compressed natural gas

CFC   Chlorofluorocarbon

NO_2 Nitrogen dioxide

O_3 Ozone

PAN   Peroxyacetyl nitrate

PM   Particulate matter

VOCS   Volatile Organic Compounds

Glossary

Air Pollution   The presence of substances in the atmosphere, particularly those that do not occur naturally. These substances are generally contaminants that substantially alter or degrade the quality of the atmosphere. The term is often used to identify undesirable substances produced by human activity, that is, anthropogenic air pollution. Air pollution usually designates the collection of substances that adversely affects human health, animals, and plants; deteriorates structures; interferes with commerce; or interferes with the enjoyment of life.

Atmosphere   A gaseous envelope gravitationally bound to a celestial body (e.g., a planet, its satellite, or a star).

Carbon monoxide   A colorless, odorless, very toxic gas; formula CO, molecular weight 28.

Carbon   Colorless gas, formula CO_2, molecular weight 44; the

dioxide   fourth most abundant gas in dry air.

Ecosystem   An interactive system that includes the organisms of a natural community association together with their abiotic physical, chemical, and geochemical environment.

Environmental policy   A policy initiative aimed at addressing environmental problems and challenges.

Fog   Water droplets suspended in the atmosphere in the vicinity the earth's surface that affect visibility.

Global warming   increase in the average temperature of the earth's surface.

Hydrocarbons   Strictly speaking, organic molecules consisting of just carbon and hydrogen; often loosely applied also to derivatives of hydrocarbons containing oxygen, halogens, etc. The atmospheric burden of hydrocarbons is provided from both natural and anthropogenic emissions.

Los Angeles Photochemical Smog   Type of air pollution characterized by high levels of ozone and low visibility, typically found in cities located in a valley (e.g., Los Angeles, Denver, Mexico City).

London Photochemical Smog   Deadly mixture of smoke and fog peaking in the mid twentieth century in large cities.

Nitrogen Oxides   Family of compounds in which nitrogen is bound to oxygen.

Ozone   A nearly colorless gas, formula O_3, molecular weight 48, that appears blue in the condensed phase or at high concentration, with a characteristic odor like that of weak chlorine.

It is formed in the reaction between atomic oxygen and molecular oxygen

Oxidants   Substance capable of causing oxidation of, for example, an atmospheric species.

Particulates   The term for solid or

liquid particles found in the air.

Photochemical Smog

Air contaminated with ozone, nitrogen oxides, and hydrocarbons, with or without natural fog being present.

In the presence of sunlight, hydrocarbons and [NO_x] are involved in a complex series of chemical reactions that eventually creates ozone and other oxidants as secondary pollutants. However, ozone is also destroyed by NO_x. Photochemical air pollution levels are generally proportional to concentrations of nitrogen oxides and hydrocarbons; they also increase with strong solar intensity and high ambient temperatures, which increase biogenic volatile organic emissions to the atmosphere from vegetation. The pollutant levels are inversely proportional to wind speed and inversion height.

Pollutants (pollution)

Unwanted chemicals or other materials found in the air. Pollutants can harm health, the environment and property. Many air pollutants occur as gases or vapors, but some are very tiny solid particles: dust, smoke, or soot.

Point pollution polluted water from a defined point. It can be collected as industrial or municipal wastewater and treated by what is often called end-of-pipe technology (environmental technology).

Pollution control

The addition of processes, practices, materials, products or energy to waste streams to reduce the risk posed by pollutants and waste before their release to the environment.

Pollution prevention

The use of processes, practices, materials, products, substances or energy that avoid or minimize the creation of pollutants and waste, and reduce the overall risk to human health or the environment

Public health

The health or physical well-being of a whole community.

Secondary pollutants

Pollutants that are formed in the atmosphere as a result of chemical reactions. Secondary pollutants are often photochemical oxidants such as ozone or nitrogen dioxide, or components of acid rain such as sulfuric acid or nitric acid.

Smoke

Foreign particulate matter in the atmosphere resulting from combustion processes

Threatened species

Species of flora or fauna likely to become endangered within the foreseeable future.

Toxic emissions

Poisonous chemicals discharged to

air, water, or land.

Toxic waste

Garbage or waste that can injure, poison, or harm living things, and is sometimes life-threatening.

Visibility

A measurement of the ability to see and identify objects at different distances. Visibility reduction from air pollution is often due to the presence of sulfur and nitrogen oxides, as well as particulate matter.

Volatile

Evaporating or vaporizing readily under normal conditions; having a low boiling point

1. Introduction

Photochemical Smog is a particular health danger in some of the world's sunniest and most populated cities. This term used during the 1950s. The energy in the sunlight converts the pollutants into other toxic chemicals.

Photochemical smog is therefore considered to be a problem of modern industrialization. It is present in all modern world, it is more common in cities with sunny, warm, dry climates and a large number of motor vehicles. It travels with the wind, it can affect sparsely populated areas as well.

Smog is derived from coal emissions, vehicular emissions, industrial emissions, forest and agricultural fires and photochemical reactions of these emissions.

Photochemical smog is a major contributor to air pollution. The word smog was formed as a mixture of smoke and fog and was used to describe air pollution produced from the burning of fossil fuels, which released smoke   and   sulfur   dioxide, vehicular emission from internal combustion engines and industrial fumes that react in the atmosphere with sunlight to form secondary pollutants that also combine with the primary emissions to form photochemical smog.

Photochemical smog is a mixture of pollutants that are formed when nitrogen oxides and volatile organic compounds (VOCs) react to sunlight, creating a brown haze above and around our cities. It tends to occur more often in summer, because that is when we have the most sunlight.

Nineteenth and 20th century London was particularly well-known for this type of air pollution. The Great Smog of 1952 was identified as the cause of over 4,000 deaths in London. This kind of visible air pollution is composed of nitrogen oxides, ozone, sulfur oxides, smoke or particulates among others (less visible pollutants include carbon monoxide, CFCs and radioactive sources).

Photochemical smog is a type of secondary pollutant that occurs when the chemicals given off react with sunlight in the atmosphere.

When nitrous oxides and VOCs interact with sunlight, secondary pollutants are formed, such as ozone and peroxyacetyl nitrate. These secondary pollutants are what we have been calling photochemical smog.

Photochemical smog is composed of primary and secondary pollutants. Primary and secondary pollutants in photochemical smog are highly reactive. These oxidizing compounds have been linked to a variety of negative health results.

The link between automotive exhaust and photochemical smog was discovered in the 1950s by Arie Haagen-Smit.

2. Overview

Smog can form in almost any climate where industries or cities release large amounts of air pollution, such as smoke or gases. However, it is worse during periods of warmer, sunnier weather when the upper air is warm enough to inhibit vertical circulation. It is especially prevalent in geologic basins encircled by hills or mountains. It often stays for an extended period of time over densely populated cities or urban areas, and can build up to dangerous levels.

The atmospheric pollution levels of Beijing, Delhi, Mexico City, Washington, Newyork, Tokyo and other cities are increased by inversion that traps pollution close to the ground. It is usually highly toxic to humans and can cause severe sickness, shortened life or death.

In Indian cities like Delhi, smog severity is often aggravated by stubble burning in neighboring agricultural areas. In fact, most major cities have problems with smog and air pollution.

The driving force for the design and implementation of emission control strategies aimed at improving air quality has been the protection of the health of the population in urban centers.

The severity of smog is often measured using automated optical instruments such as Nephelometers, as haze is associated with visibility and traffic control in ports.

3.   Chemistry of Photochemical smog

A primary pollutant is an air pollutant emitted directly from a source. A secondary pollutant is not directly emitted as such, but forms when other pollutants (primary pollutants) react in the atmosphere.

Photochemical smog forms when primary pollutants react with ultraviolet light to create a variety of toxic and reactive compounds.

The two major primary pollutants, nitrogen oxides and VOCs, combine to change in sunlight in a series of chemical reactions, outlined below, to create what are known as secondary pollutants.

The secondary pollutant that causes the most concern is the ozone that forms at ground level. While ozone is produced naturally in the upper atmosphere, it is a dangerous substance when found at ground level. Many other hazardous substances are also formed, such as peroxyacetyl nitrate (PAN).

Products like ozone, aldehydes, and peroxyacetyl nitrates are called secondary pollutants, This mixture composed of air pollutants may include the following:

- Aldehydes
- Nitrogen oxides, particularly nitric oxide and nitrogen dioxide
- Peroxyacetyl nitrates
- Tropospheric ozone
- Volatile organic compounds

Photochemical smog contains a number of chemicals which are reputed to cause damage to health, materials and vegetation. It is the result of a complex sequence of chemical reactions, initiated by the action of sunlight on nitrogen dioxide and fed by non-methane hydrocarbons and air.

Nitrogen dioxide (NO_2) can be broken down by sunlight to form nitric oxide (NO) and an oxygen radical (O):

1) NO_2 + sunlight → NO + O

Oxygen radicals can then react with atmospheric oxygen (O_2) to form ozone (O_3):

2) O + O_2 → O_3

Ozone is consumed by nitric oxide to produce Nitrogen dioxide and oxygen:

3) O_3 + NO → NO_2 + O_2

Harmful products, such as PAN, are produced by reactions of nitrogen dioxide with various hydrocarbons (R), which are compounds made from carbon, hydrogen and other substances:

4) NO_2 + R → products such as PAN

The main source of these hydrocarbons is the VOCs. Similarly, oxygenated organic and inorganic compounds (RO_x) react with nitric oxide to produce more nitrogen oxides:

5) $NO + RO_x \rightarrow NO_2 +$ other products

The significance of the presence of the VOCs in these last two reactions is paramount.

Ozone is normally consumed by nitric oxide, as in reaction 3. However, when VOCs are present, nitric oxide and nitrogen dioxide are consumed as in reactions 4 and 5, allowing the buildup of ground level ozone.

Harmful products, such as PAN, are produced by reactions of nitrogen dioxide with various hydrocarbons (R), which are compounds made from carbon, hydrogen and other substances: 4) $NO2 + R$, products such as PAN, Ozone is irritate the lungs which is formed when hydrocarbons (HC) and nitrogen oxides (NOx) combine in the presence of sunlight; nitrogen dioxide (NO2), which is formed as nitric oxide (NO) combines with oxygen in the air; and acid rain, which is formed when sulfur dioxide or nitrogen oxides react with water. All of these harsh chemicals are usually highly reactive and oxidizing,

The two major primary pollutants, nitrogen oxides and VOCs, combine in a series of chemical reactions to create secondary pollutants, which are dangerous when detected in our atmosphere and at ground level.

The two most dominant toxic components produced in photochemical smog are ozone and peroxyacetyl nitrate.

Ozone produced naturally in an unpolluted environment, the interaction with VOCs prevents nitrogen oxide after being transformed to nitrogen dioxide from consuming ozone.

This leads to toxic, harmful levels of ozone in the immediate environment. Smog can also affect areas of the country that are sunny less frequently.

4. Causes and effects of Photochemical smog

I. Natural Causes

An erupting volcano emitted high levels of sulphur dioxide along with a large quantity of particulate matter. However, the smog created as a result of a volcanic eruption as a natural occurrence.

The radiocarbon content of some plant life has been linked to the distribution of smog in some areas. For example, the creosote bush in the Los Angeles area has been shown to have an effect on smog distribution that is more than fossil fuel combustion alone.

II. Anthropogenic Causes

Primary pollutants, which include nitrogen oxides and volatile organic compounds, are introduced into the atmosphere via vehicular emissions and industrial processes.

Excessive Fossils fuels used to fulfilled countries electricity demand, allow us for public/private transportation, and are the means for powering factories that manufacture everything.

Primary pollutants, introduced into the atmosphere through automobile emissions and industrial processes. Ultraviolet light can split nitrogen dioxide into nitric oxide and monatomic oxygen; this monatomic oxygen can then react with oxygen gas to form ozone.

The ozone in smog also affects plants growth and can cause widespread damage to crops and forest, and the haze reduces visibility. It can cause serious problems with our lungs and vision.

Peroxyacetyl nitrate is one of the chemicals that is responsible for damaging lung tissue, and photochemical smog forms plenty of it.

It has also been estimated that more than 100 million people around the world live in areas with ozone levels above the established standards for health safety.

a) Coal

Coal fires, used to heat individual buildings or in a power-producing plant, can emit significant clouds of smoke that contributes to smog.

b) Transportation emissions

The major culprits from transportation sources are carbon monoxide (CO), nitrogen oxides (NO and NO_x), volatile organic compounds, sulfur dioxide, and hydrocarbons.

Trucks, buses and automobiles also contribute. Airborne by-products from vehicle exhaust systems cause air pollution and are a major ingredient in the creation of smog in cities.

Hydrocarbons are the main components of petroleum fuels such as gasoline and diesel fuel. These molecules react with sunlight, heat, ammonia, moisture, and other compounds to form the noxious vapors, ground level ozone, and particles that comprise smog.

Photochemical smog damage to ecological systems caused by air pollution is an issue that is particularly worrisome if one considers the current gnawing population and urbanization trends.

5. Effects of Photochemical smog on Human Health

Photochemical smog can have an effect on the whole environment like people's health and on various physical materials. Smog is a serious problem in many cities and continues to harm human health.

Ground level ozone, sulfur dioxide, nitrogen dioxide and carbon monoxide are especially harmful for Infants, senior citizens, children, and people with heart and lung conditions such as emphysema, bronchitis, and asthma.

Toxic levels of ozone created by photochemical smog can damage lung tissue and lower the immune system. The sulfuric acid created by industrial smog has caused thousands of fatalities beginning with the Industrial Revolution and continuing until today, also cause breathing problems and skin irritations, and can even corrode buildings.

According to a research study even a very small 5 µg amount of PM2.5 exposure was associated with an increase (18%) in risk of a low birth weight at delivery, and this relationship held even below the current accepted safe levels.

It can inflame breathing passages, pneumonia, inflammation of pulmonary tissues, heart attacks, decrease the lungs working capacity, cause shortness of breath, eye and nose irritation, lung cancer, increased asthma-related symptoms, fatigue, heart palpitations, and even premature ageing of the lungs, bronchitis, pain when inhaling deeply, wheezing, and coughing. it dries out the protective membranes of the nose and throat and interferes with the body's ability to fight infection, increasing susceptibility to illness and death.

According to the leading Lung Association's around the world, human lungs can get permanently damaged due to prolonged exposure to air pollution and smog. Respiratory diseases and breathing shortness/ problems are often increase during periods when ozone levels are high. There is a lack of knowledge on the long-term effects of air pollution exposure and the origin of asthma.

The U.S. EPA has developed an Air Quality Index to help explain air pollution levels to the general public. 8 hour average ozone concentrations of 85 to 104 ppb are described as Unhealthy for Sensitive Groups, 105 ppb to 124 ppb as unhealthy and 125 ppb to 404 ppb as very unhealthy. The very unhealthy range for some other pollutants are: 355 μg m^{-3} - 424 μg m^{-3} for PM10; 15.5 ppm - 30.4ppm for CO and 0.65 ppm - 1.24 ppm for NO_2.

The Ontario Medical Association announced that smog is responsible for an estimated 9,500 premature deaths in the province each year.

American Cancer Society study found that cumulative exposure also increases the likelihood of premature death from a respiratory disease, implying the 8-hour standard may be insufficient.

A study examining women who had babies with birth defects between 1997 and 2006, and 849 women who had healthy babies, found that smog in the San Joaquin Valleyarea of California was linked to two types of neural tube defects: spina bifida (a condition involving, among other manifestations, certain malformations of the spinal column), and anencephaly (the underdevelopment or absence of part or all of the brain, which if not fatal usually results in profound impairment).

Avoid exercising near places with heavy traffic, especially during peak hour. Also, avoid outdoor activities when smog levels are high, especially during the afternoon. If you have a heart/lung condition, consult your doctor about ways to protect your health from smog.

Table: Photochemical smog Pollutants and their Health effects

Pollutants	Effects
Nitrogen oxides	• Can contribute to problems with heart and lungs • Links to decreased resistance to infection
Volatile organic compounds (VOCs)	• Eye irritation • Respiratory problems • Some compounds are carcinogens
Ozone	• Coughing and wheezing • eye irritation • respiratory problems (particularly for conditions such as asthma)
Peroxyacetyl nitrate (PAN)	• Eye irritation • Respiratory problems

6. Effect of photochemical smog on Plants

Chemicals such as nitrogen oxides, ozone and Peroxyacetyl nitrate (PAN) can have harmful effects on plants. These substances can reduce or even stop growth in plants by reducing photosynthesis. Ozone, even in small quantities, can achieve this, but PAN is even more toxic to plants than ozone.

7. Effect of photochemical smog on Nonliving Beings

Photochemical smog can damage various Physical materials. It can cause the cracking of rubber, the reduction in tensile strength of textiles, fading of dyed fibres and cracking of paint and potential to damage artworks and books of museums and libraries.

8. Effect of photochemical smog in reference to Delhi, India

Delhi is the most polluted city in the world and according to one estimate, air pollution causes the death of about 10,500 people in Delhi every year. During 2013-14, peak levels of fine particulate matter (PM) in Delhi increased by about 44%, primarily due to high vehicular and industrial emissions, construction work and crop burning in adjoining states.

Delhi has the highest level of the airborne particulate matter, PM2.5 considered most harmful to health, with 153 micrograms. Rising air pollution level has significantly increased lung-related ailments (especially asthma and lung cancer) among Delhi's children and women.

The dense smog in Delhi during winter season results in major air and rail traffic disruptions every year.

According to Indian meteorologists, the average maximum temperature in Delhi during winters has declined notably since 1998 due to rising air pollution.

Environmentalists have criticised the Delhi government for not doing enough to curb air pollution and to inform people about air quality issues. Most of Delhi's residents are unaware of alarming levels of air pollution in the city and the health risks associated with it. Since the mid-1990s, Delhi has undertaken some measures to curb air pollution – Delhi has the third highest quantity of trees among Indian cities and the Delhi Transport Corporation operates the world's largest fleet of environmentally friendly compressed natural gas (CNG) buses.

In 1996, the Centre for Science and Environment (CSE) started a public interest litigation in the Supreme Court of India that ordered the conversion of Delhi's fleet of buses and taxis to run on CNG and banned the use of leaded petrol in 1998.

In 2003, Delhi won the United States Department of Energy's first Clean Cities International Partner of the Year award for its bold efforts to curb air pollution and support alternative fuel initiatives. The Delhi Metro has also been credited for significantly reducing air pollutants in the city.

According to CSE and System of Air Quality Weather Forecasting and Research (SAFAR), burning of agricultural waste in nearby Punjab, Haryana and Uttar Pradesh regions results in severe intensification of smog over Delhi.

The state government of adjoining Uttar Pradesh is considering imposing a ban on crop burning to reduce pollution in Delhi NCR and an environmental panel has appealed to India's Supreme Court to impose a 30% cess on diesel cars.

9. Method of Control/reduction to photochemical smog

Although, air pollution and smog are something that cannot be stopped, but it can be surely controlled if appropriate steps are taken at an individual level. The degradation of air quality is no longer just a local problem; it is beginning to acquire regional and even global attentions, and to affect large portions of the world.

Photochemical smog is a serious problem for urban areas that are in a geographical location that contributes to its formation. The most effective way of reducing the amount of secondary pollutants created in the air is to reduce emissions of both primary pollutants. Cumulative efforts of community involved can help in reducing the impact of smog on area.

A serious attempt to reduce personal pollutant output. Walk or bike to places that are short distances away instead of driving. Arrange to be involved in a local car pool or use public transportation instead of driving and fill up with gasoline after dark. Turn off lights, air conditioning and electronic devices when not in use. Reduce electricity usage. Burning coal is the primary source of electrical production, so avoid excessively using electricity.

A catalytic converter fitted to a car's exhaust system will convert much of the nitric oxide from the engine exhaust gases to nitrogen and oxygen. Lowering the levels of nitrogen oxides is by Catalytic reduction, which is used in industry and in motor vehicles. Using less air in combustion can reduce emissions of nitrogen oxides.

Temperature also has an effect on emissions—the lower the temperature of combustion, the lower the production of nitrogen oxides. Temperatures can be lowered by using processes such as two-stage combustion and flue gas recirculation, water injection, or by modifying the design of the burner.

There are various ways to reduce VOC emissions from motor vehicles. These include the use of liquefied petroleum gas (LPG) or compressed natural gas (CNG) rather than petrol, decreasing distances vehicles travel by using other modes of transport, such as buses and bikes, and implementing various engine and emission controls now being developed by manufacturers.

Some other controlled strategies are:

• Keep motor vehicle regularly serviced and the tyres inflated to the manufacturer's specifications, ensure the car is running efficiently and not emitting excessive pollutants.

• Replace old car with a fuel-efficient, low emission car. Check out the fuel consumption label, which now has to be displayed on new cars.

• Instead of using a car riding use buses, trains or public transportation, whenever possible.

• Use energy efficient appliances.

• If renovating or building, use energy-efficient designs and materials.

• Should used Energy Star logo on electronics product; or the Energy Rating on the air conditioner, clothes dryer, washing machine, dishwasher and fridge to purchase.

• Turn off unnecessary electrical appliances at the power point wherever possible.

• Should use power generated from clean, renewable energy sources.

• Limit your wood fires at home. Wear warmer clothes as your first action to keep warm.

• Should use solar and other Renewable energy power.

• Schools can get involved with innovative environment learning program for schools, where students can become pollution watchdogs in their local areas.

10. References

1. Colbeck, I & Mackenzie, AR 1994, Air pollution by photochemical oxidants, Air Quality Monographs Vol 1, Elsevier, Netherlands.

2. White, V 1998, Air emissions inventory for the Adelaide airshed 1995, Environment Protection Agency, South Australia.

3. http://www.pmfias.com/2015/08/Smog-Photochemical-smog.html

4. Smog — Who Does It Hurt? What You Need to Know About Ozone and Your Health (EPA-452/K-99-001). United States Environmental Protection Agency. July 1999

5. The Regional Transport of Ozone: New EPA Rulemaking on Nitrogen Oxide Emissions (EPA-456/F-98-006)

6. https://en.wikipedia.org/wiki/Smog

ABOUT THE AUTHOR

Dr. Hemant Pathak held positions as Assistant Professor in the department of chemistry, Govt. Indira Gandhi Engineering College, Sagar, MP, India. He had extensive experience in teaching, research and administrative management.

Dr. Pathak received his Ph.D. degree in chemistry from Dr. Hari Singh Gour Central University, Sagar, India and M.Sc. Gold medalist from Jiwaji University, Gwalior. He has published 25 books and more than 50 research papers in reputed International and National journals and received several awards. He is a member of editorial boards and reviewer boards of several international journals and societies. His area of specialization includes Engineering Chemistry, Energy audits and Environmental Pollution management.

www.ingramcontent.com/pod-product-compliance
Lightning Source LLC
Chambersburg PA
CBHW061233180526
45170CB00003B/1272